ПРОЛЕТАРИИ ВСЕХ СТРАН, СОЕДИНЯЙТЕСЬ!

ПРАВДА ВОСТОКА

Газета издается с апреля 1917 года

ОРГАН ЦЕНТРАЛЬНОГО КОМИТЕТА КОМПАРТИИ УЗБЕКИСТАНА И СОВЕТА МИНИСТРОВ УЗ.ССР

| №127 (15072) | Среда, 8 июня 1966 года | Цена 2 коп |

The Aral Sea (pronounced *ar-uhl see*) is situated between Kazakhstan in the north and Uzbekistan in the south. It appeared after the last Ice Age, when glaciers in the Pamir Mountains and the Tian Shan Mountains began to melt. Two large rivers, the Amu Darya and the Syr Darya, carried down the waters and formed the Aral Sea. It was so enormous that the locals called it a sea even though it is really a vast lake. Before the 1960s, it was the fourth-largest lake in the world. For centuries, people loved living around it—the water was potable, helped gardens grow, and provided fish to catch. Today all of that is history because of a few irrigation projects that prioritized growing cotton over the environment.

The Changing Profile of the Aral Sea

SYR DARYA

MOYNAQ

AMU DARYA

1960

These maps show the progressive shrinking of the Aral Sea over a span of 60 years, highlighting significant reductions in water volume and surface area.

Big gratitude to the wonderful people who played key roles in publishing this story—Zarrina Talipova, Kasim-oka, Ariel Richardson, Jennifer Tolo Pierce, Sarah Cameron, and Hudson Library and Historical Society.

TO THE PEOPLE OF ARAL.

Library of Congress Cataloging-in-Publication Data available.

ISBN 978-1-7972-2459-6

Manufactured in China.

Design by Jennifer Tolo Pierce.
Typeset in Lucida Sans Typewriter and custom font by Dinara Mirtalipova.
The illustrations in this book were rendered in the soft pastel stenciling technique.

10 9 8 7 6 5 4 3 2 1

Chronicle Books LLC
680 Second Street
San Francisco, California 94107
www.chroniclekids.com

The Vanishing Sea

The Tale of How the Aral Sea Became the Aral Desert

Dinara Mirtalipova

CHRONICLE BOOKS
SAN FRANCISCO

A long, long time ago,
there was a lake so vast
that the People called her
MOTHER SEA.

The People were small and **HUNGRY**. They praised the Sea and asked her to feed them.

Like a caring mother, the Sea gave them
her biggest fish . . .

and taught them
how to fish.

So the People

were never
hungry again.

The People grew larger and cried,
"Mother Sea, we are THIRSTY!"

Like a caring mother,
the Sea gave them water
to drink.

Villages formed.

Cities were built.
The People grew even LARGER.

The People still cried,

"Mother Sea,
we want to get RICHER!"

Like a selfless mother,
the Sea gave them ALL
of her resources.

When the people returned
to ask for MORE,

they realized there was
very little left

of Mother Sea.

Soon they had no fish to catch . . .

and no water to drink.

Can the People save Mother Sea?

Author's Note

I was born in the 1980s in Tashkent, Uzbek Soviet Socialist Republic (now Uzbekistan). It was the Soviet Era. I recall hearing about the shrinking Aral waters as a child, but I don't remember anyone doing anything about it. It was the era of perestroika (a Soviet plan for restructuring) and a time of uncertainty. Like many families, mine has since migrated to another country. As I morphed from a teenager into an adult and then into a parent, I kept coming across articles about the disappearance of the Aral Sea. Every time, it scratched my heart a little deeper. In the summer of 2023, I decided to see it with my own eyes.

Me, on my trip to the Aral Sea, 2023

The journey was rough. Only experienced local caravanners know the way, as the majority of the path is an endless, empty seabed. I arrived in Nukus, Uzbekistan, and hired a driver named Kasim-oka to take me first to the town of Moynaq—once a port on the Aral Sea—and then to the southern edge of what's left of the Aral Sea's waters. Right after we left Moynaq we lost all forms of communication (including GPS!) for the remainder of our journey. We packed enough water and food to last for a few days. As we drove on the bumpy seabed surface, we made occasional stops to take photos.

This journey was life-changing!

Yurt camp

Moynaq

Nukus

Tashkent

What seemed like big mountains—or, as locals call them, canyons—were in fact once underwater ridges. Only some 50 years ago, they were covered with water.

Moynaq in the 1960s was a booming town with one of the largest fish-canning factories in the former Soviet Union. Today it's a ghost town with remnants of fishing boats spread across the deserted seabed like rusty skeletons.

What caused this massive body of water to disappear? The answer is greed, cotton production, and the Soviet five-year economic plan. In the 1950s, the Soviet Union sought to boost cotton production to rapidly catch up with the capitalist West. Construction began on a large irrigation canal, the Karakum Canal, through Turkmenistan, which siphoned off water from the rivers that fed the sea, the Amu Darya and the Syr Darya.

Cotton is a delicate plant that requires lots of water. The territory of Uzbekistan was perfect for raising it. And as the highest-quality cotton can only be collected by hand, most children in Uzbekistan were taken from their schools and sent to the fields to pick it. My mom and dad were among them. Free child labor was considered a service to the country.

Yurt camp, Aral Sea, 2023

Unfortunately, a lot of water leached into the porous and muddy soil, and much of it never made it to the fields. And soon the water ran out.

As a result of this failed irrigation system, the world's fourth-largest lake was drained from the surface of the earth over the course of just a few decades. The region I visited is the homeland of the Qaraqalpaqs (pronounced *kah-rah-kall-pahks*), a Turkic minority group in Uzbekistan. Today the Qaraqalpaqs struggle to sustain basic living conditions. Drinking water is scarce, and the soil is full of toxins.

My proud grandparents visiting Mom in a cotton field, Aral Sea, 1963

Mom, Aral Sea, 1963

A Brief History

Syr Darya

Ancient Khwarazm

The story of the Aral Sea begins about 17,000 years ago, when the glaciers from the last Ice Age began to melt in Ancient Khwarazm. People called the new body of water a sea because it was so large—but in fact, the Aral Sea is a closed basin lake. It became the fourth-largest lake in the world.

Amu Darya

17,000 years ago–2nd cent. CE

305 CE–995 CE

1077 CE–1231 CE

Afrighid Dynasty

Under the Afrighid dynasty, Khwarazm transitioned from Zoroastrianism to Islam and evolved into a key cultural and commercial center on the Silk Road.

Khwarazmian Kingdom

The Khwarazmian Kingdom flourished and achieved significance under the rule of Khwarazmshahs (shahs, or kings) who expanded its territory and fostered advancements in science and the arts. This era of prosperity ended with the Mongol invasion by Genghis Khan, triggered by a diplomatic dispute that devastated the kingdom.

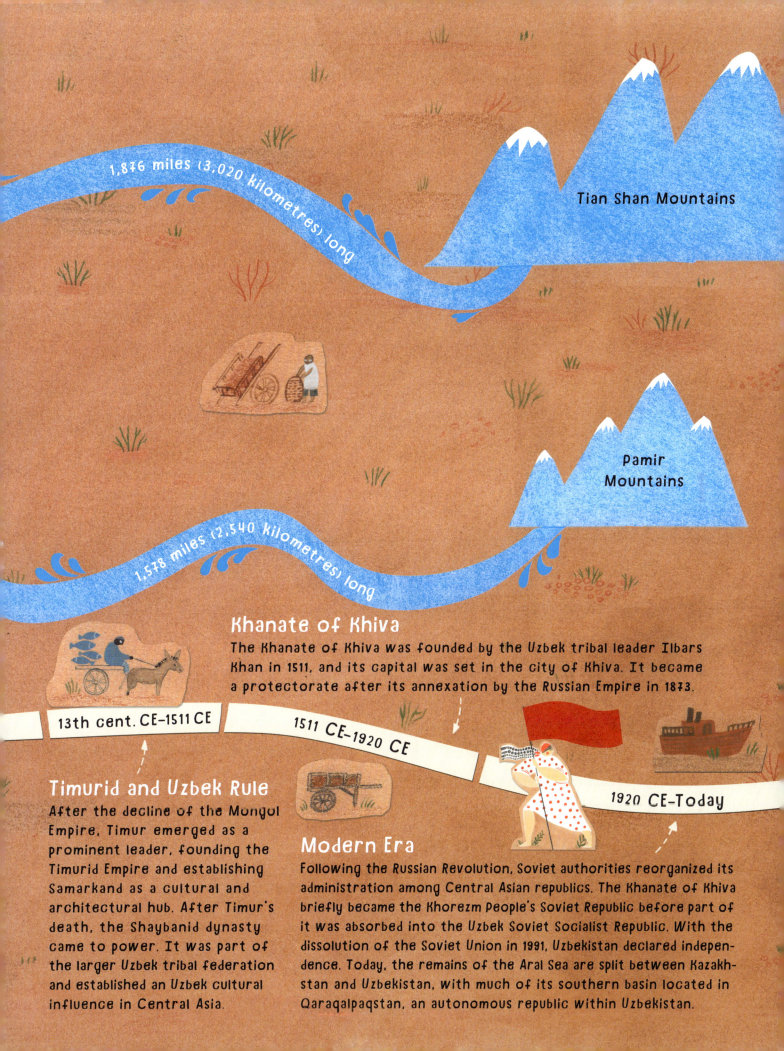

1,876 miles (3,020 kilometres) long

Tian Shan Mountains

Pamir Mountains

1,578 miles (2,540 kilometres) long

Khanate of Khiva

The Khanate of Khiva was founded by the Uzbek tribal leader Ilbars Khan in 1511, and its capital was set in the city of Khiva. It became a protectorate after its annexation by the Russian Empire in 1873.

13th cent. CE–1511 CE

1511 CE–1920 CE

1920 CE–Today

Timurid and Uzbek Rule

After the decline of the Mongol Empire, Timur emerged as a prominent leader, founding the Timurid Empire and establishing Samarkand as a cultural and architectural hub. After Timur's death, the Shaybanid dynasty came to power. It was part of the larger Uzbek tribal federation and established an Uzbek cultural influence in Central Asia.

Modern Era

Following the Russian Revolution, Soviet authorities reorganized its administration among Central Asian republics. The Khanate of Khiva briefly became the Khorezm People's Soviet Republic before part of it was absorbed into the Uzbek Soviet Socialist Republic. With the dissolution of the Soviet Union in 1991, Uzbekistan declared independence. Today, the remains of the Aral Sea are split between Kazakhstan and Uzbekistan, with much of its southern basin located in Qaraqalpaqstan, an autonomous republic within Uzbekistan.

Local and Global Ecological Impact

The shrinkage of the Aral Sea has had a great impact on the Qaraqalpaqs, who have long depended on its resources, as well as on other communities in the region, including the Kazakh (*kuh-zahk*) people along the northern shore. For the Qaraqalpaqs, there's no longer clean water close by—it has to be delivered to their villages by the bucketful. There are also no more fish left for them to catch.

The overuse of pesticides in the region combined with the high salinity of the evaporated Aral waters and the resulting dust are key environmental contributors to serious health conditions for locals, including cancers, anemia, thyroid problems, and weak immune systems. There is evidence that toxic storms lift heavy metal particles from the area and carry them across Eurasia, impacting people as far away as Greenland and Norway. In addition, there has been an incredible loss of biodiversity, from birds and fish to flora and other fauna.

QARAQALPAQS: a Turkic minority group near the Aral Sea, pronounced with a throaty K.
/KAH-RAH-KALL-PAHKS/

Literal translation: black hats. QARA: black. QALPAQS: hats.

Before the irrigation project in the 1960s, the Aral Sea was home to a thriving ecosystem that supported many native freshwater fish species and a bustling fishing industry.

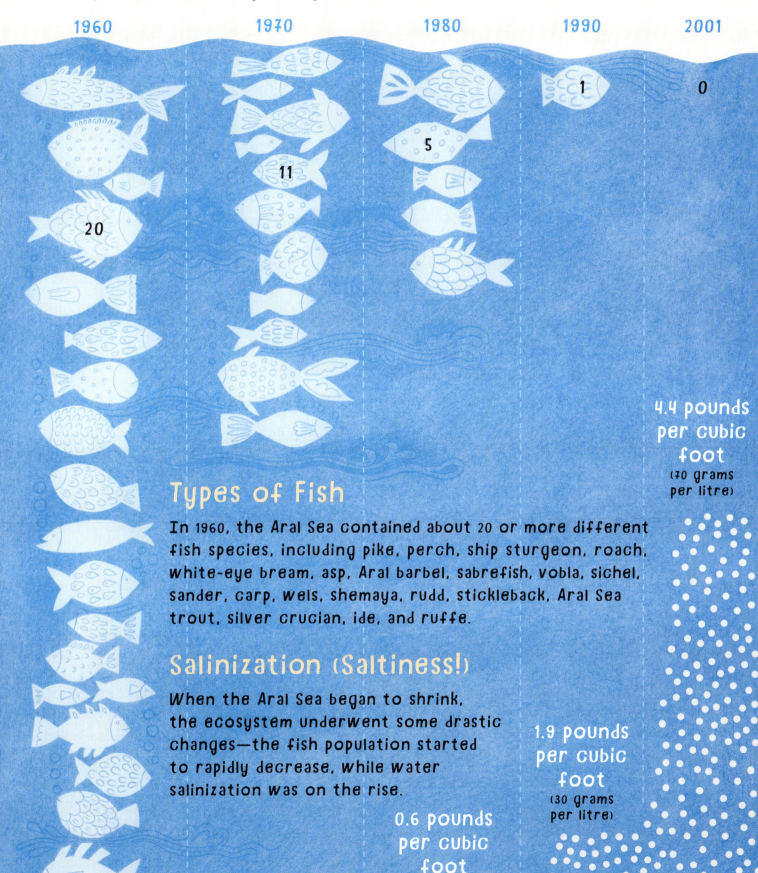

1960   1970   1980   1990   2001

20   11   5   1   0

Types of Fish

In 1960, the Aral Sea contained about 20 or more different fish species, including pike, perch, ship sturgeon, roach, white-eye bream, asp, Aral barbel, sabrefish, vobla, sichel, sander, carp, wels, shemaya, rudd, stickleback, Aral Sea trout, silver crucian, ide, and ruffe.

Salinization (Saltiness!)

When the Aral Sea began to shrink, the ecosystem underwent some drastic changes—the fish population started to rapidly decrease, while water salinization was on the rise.

4.4 pounds per cubic foot
(70 grams per litre)

1.9 pounds per cubic foot
(30 grams per litre)

0.6 pounds per cubic foot
(10 grams per litre)

WANTED

MISSING SEA

last seen in 1960

If found, please return to the People of Aral